So You Want To Build the Naked UFO/Plasma Active Craft

Set Plasma Free

chipmunkapublishing
the mental health publisher

Set Plasma Free

All rights reserved, no part of this publication may be reproduced by any means, electronic, mechanical photocopying, documentary, film or in any other format without prior written permission of the publisher.

>Published by
Chipmunkapublishing
PO Box 6872
Brentwood
Essex CM13 1ZT
United Kingdom

http://www.chipmunkapublishing.com

Copyright © Set Plasma Free 2010

Chipmunkapublishing gratefully acknowledge the support of Arts Council England.

So You Want To Build the Naked UFO/Plasma Active Craft

Author Biography

Set Plasma Free knows what it is like to be bottom of the class, in every subject, and to leave school at the age of 16, to wash dishes and make beds. All those years ago Dyslexia was not a known disability neither was Asperger's Syndrome. In Set Plasma Free's case his brain could see things that other people couldn't or did not want to see.

However, Set Plasma Free has the ability to communicate with his voice which he believes is a gift from God for helping him to rise above all the ridicule that has been thrown his way over all these years.

Thanks to Chipmunka Publishing for giving people like Richard an opportunity to write and publish his works.

Set Plasma Free

So You Want To Build the Naked UFO/Plasma Active Craft

The draft sketch is a cheap and cheerful test rig for the guts of the plasma active craft. It also doubles as a test rig to prove the concept of hubless high speed flywheel rotor.

In these innocent words is a hidden warning that you ignore at your peril. Alien power source could well be a device that converts the very air we all breathe and the very water that we drink directly into energy/power.

A metamorphism not seen since the young Jesus of Nazareth allegedly turned the water into wine. Alien power sources could trace its lineage to the mythical arc of the covenant – a revered source of energy/power used by the ancient people of Israel. In our clichéd ridden times: "those who play with fire will get their fingers burn" "electricity and water do not mix", "fools rush in where angels dare to tread" – do you all now get my drift? This is a health and safety matter in a league of its own.

So you still want to build the naked UFO/Plasma Active craft?

Could I suggest that you make a fly on the wall documentary and as you do so, not only will this be a record of your progress but by sleight of hand could end up as a scrutinised debt of obligation, a financial instrument by any other name.

To help you on your way, in the appendices are a series of philosophical scientific notes to support my hypotheses:

Gone With A Bang
The Directors Cut
A World Run on Water and the Air We Breath Should Not Be Subject Too Hot To Handle
If At First You Don't Succeed Try Try again

Earlier papers also included in the indices

Sponsorship (Blue Sky Research and development into speculative projects copyright 25.08.2004)
Sponsorship2 (copyright 28.09.2004)

Knowledge must, in the hands of some, be considered a two edged sword for they will fight tooth and nail to get their message across. Each updated concept is considered a small incremental/evolutionary step on the path to acceptance/validation of a hypothesis; all these steps cascade down so fast to culminate as one revolutionary idea. Mankind must surely benefit from his ability to harness an alien power source, just as he has done by harnessing the power of the atom – when a runaway nuclear chain reaction converted mass directly into energy, in the birth of the first atomic bomb and a few years later in the plutonium hydrogen bomb. Today we take for granted the heat and power generated by controlled nuclear chain reaction of the uranium atom leading to energy/power from fission.

On a rhetorical question, that needs to be asked again, does Mankind stand on the threshold of sustainable nuclear fusion? Some who are of optimistic inclination will say yes, others of a pessimistic point of view will say no.

Returning to the draft sketch, I can still hear my father's oath as I spilt his valuable coffee over one of those drawings with white lines on a blue background – a drawing office technology which has been by-passed by the advent of computer generated 2D/3D solid modelling and rapid prototyping – a "blue print".

Mein Gott, Donner and Blitzen.

My father was an unassuming man whose love for family was many times physically illustrated as he would make a point of having spontaneous family bear hugs, which invariably included Ralph, his faithful big silly and over playful German shepherd.

War in cold calculating terms is a matter of statistics, the loss of a sibling, mother and father is a tragedy like no other. I shall not dwell on the loss of my bear hugging family and my childhood so ruthlessly snatched away.

I will however treasure the few mementoes left to me in the buried safe which survived unscarred after the inferno had finished incinerating the outhouse, where paradoxically it was first exposed and then buried under rubble where it lay undisturbed. All I have and must hold dear are the original drawings, with my father's handwritten notes, penned with his favourite fountain pen – a gift from my mother, when she learnt he had been passed over for promotion, for he was not considered a good party member.

His laboratory was now to fall into disrepair as ubiquitous buckets stood testimony for they would have to catch the water as it seeped through too many rotten joints in the roof of his wooden laboratory. Cold and damp, he made purposely slow progress on solving the stability of certain chemical compounds needed for experiments in the new big hospital. My father used this euphemism to explain the vast construction work Peenemunde rocket research establishment, thus saving us from the reality of the adult world. He was never envious of the new built wooden huts that multiplied weekly to house more and more imported workers who would build and work in the new big hospital.

Grandfather was a proud aristocrat of Prussian stock, and saw it as his duty to make available to the new regime his distant estates at Peenemunde, his only son also had his father's pride in being German, but the regime had placed him in an invidious position, for to help them through his genius as a chemist to march on with their quest for world domination was an anathema to him.

Donner and Blitzen is a German uttering, but my father would risk all and pen some of his footnotes in English – the original sketch is no exception to this rule; consequently when presented to others all became clear. My father, on that faithful day, had not uttered an oath but had had his Eureka moment.

Donner and Blitzen – Thunder and Lightening – water air and electricity, were to be used to power five pulse jets, built onto a decagon (a ten sided figure). Each jet would feed off the other until a burning ring of fire from combustion paraffin would be held round the circumference/periphery of a spinning Catherine wheel rotor. I lost my family but the bombing of Peenemunde killed off this new terror weapon before it could see the light of day.

Pulse jets can now be substituted for the use of sixteen plus air intakes, to resemble road traffic cones. Now add thirty two smaller cones which can be conjoined at their base so that this arrangement might fit snugly in a horizontal position inside each mother cone. You can see this clearly in the sketch. All the cones do is deflect the ram air as it is rammed into the mother cone at its open base. This air now exits at high pressure due to the inside of the cone compressing the air by restricting its passage to its exit point which is through a jet nozzle fixed into the top of each mother cone. This continuous exit velocity cycle meets additional ram air, which will collide with the new ram air; with the result that high pressure air now mixes with new ram air to harness heat by friction.

Hubless high speed rotor/flywheel must have copious amounts of energy put in it to bring it to its critical state, for it has to overcome surrounding air friction and also put energy in to ram the air into the mother cones and over the smaller cones which I now call Ram Vortex Combustors (RVC). Combustor indicates that a volatile gas will be ignited with the help of vast amounts of air to accelerate the combustion process. Inside diameter of the rotor will also have the drag of the slipstream alternators with their distinctive ducted fan air driven turbines to contend with. Alien power sources must have a different fuel otherwise it would not be considered alien. Water at a very high pressure, mixed with air, is now our alien combustible fuel.

Droplets of water, as in a pressure mist, will flash over to steam if they are subject to instantaneous heat above 200 o C. This then is a target temperature for the air as it is continually trapped in the RVCs, due to two contradictory forces acting in a symbiotic relationship with each other – namely, centrifugal force to throw the RVC air out and centrapetal force to hold it in. Hold in, this time it is the jet exhaust from each mother cone as high pressure steam, has a far higher specific gravity/mass than just the hot air. This jet thrust will ease the pressure on the drive current so less electrical power will be consumed to get the high-speed rotors/flywheels up to its critical phase.

Donner and Blitzen, thunder and lightning. We have the thunder as the air moves from subsonic speed as it enters the mother cones at low rpm, with the vast majority of the air that surrounds the exposed rotor passing over the outside of the mother cones, to produce a high and low pressure toroidal spinning vortex holding and insulating its hold toroidal core. 25,000 RPM and the air has gone supersonic. To reach the target speed of 45,000 RPM we must add the lightning.

The sketch shows 16 devices, eight of which are ram vortex driven turbo DC brushed, homopolar motors; the remaining eight are also ram vortex driven but this time are DC brushed homopolar generators, designed to generate a very high voltage, to ionise the air to facilitate a high current flow when the motors are back driven by the kinetic energy in the flywheel. This kinetic energy was built up by the drive motors before the power was switched off enabling them to extract the energy to act as high current generators.

Kinetic energy used in this way and coupled to the high current flow will react with ionised air to produce a very powerful Donner and Blitzen.

I can now understand my father's excitement, intrepidation, fear and great foreboding should his hypothesis fall into the wrong hands.

Achtung, Achtung: my father's handwritten notes both in English and German, the oxygen in the air to an oxidising agent, ozone, the inert gas nitrogen, which is 75% of what we breathe into a volatile explosive oxidiser of the nitrite family. Any water molecules will immediately end up as hydrogen gas and more ozone. This then was my father's legacy. Donner and Blitzen, thunder and lightning – the air we breathe and the water we drink directly into the explosive equivalent of combusted rocket fuel.

Early jet experimental engines suffered either from flame out, i.e. they just shut down, or a runaway (could not be shut down). Plasma jet thrusts from 16 mother cones, as per the sketch, will pass into the runaway phase with so much power and heat generated that, and I now quote from my father's notes, on the work done by a press-ganged member of the rocket research team, Henri-Marie Coanda, who came to the attention of the regime when he postulated in 1930 that if air could be blown over a shallow hemisphere then that hemisphere would act like an aerofoil, with the blown/sucked air flowing to the underside of the hollow hemisphere at such a rate to literally suck the hemispheric circular aerofoil straight up.

So You Want To Build the Naked UFO/Plasma Active Craft

Unconventional applications will see our eight homopolar motors used their through driveshafts to drive what looks like a dimpled cricket ball with a raised spiral spring steel seam. The inner dimpled sphere will need to rotate faster than its counterpart on the other end of the through shaft – so consequently will have a smaller overall diameter.

The raised seam will press down and engage in a track made of identical spiral spring steel, which in turn sits flush at the bottom of a machine groove designed to accept 25% of the rolling circumference of the dimpled cricket ball like sphere.

This arrangement allows for a series of looped spiral steel bands to compress the two fixed discs up against the rolling dimpled cricket ball like spheres.

Golf, yes my father would delegate me to approach my mother every time he wanted to play a round of golf. He was naturally very proud of his personally designed nine hole course.

The request was always the same, as was the answer. Would she and her children like to pile onto the old cart pulled by an ancient mule, left behind at the end of the Great War, for a fun day out? My mother would politely decline the invitation but would see to it that the staff under her control would pack us all off with a wholesome picnic, contained within a purpose woven wicker hamper.

Off we would all go then after a family bear hug.

Father would always have a plentiful supply of golf balls which he would distribute to all and sundry, including Ralph who would push, chase and pounce on them until he got bored – and would then lay in the sun as his master talked to his children about the dimples in the golf ball made it fly further and straighter then the smooth ball.

Looking back I can recall the bemused expression on the faces of friends and visitors as my father would press a golf ball into their hands, with his pet saying repeated "one day, one day".

Now I understand my father's fascination for golf balls and the need for substitution of cones for pulsed jets – he recognised that burning paraffin would never produce enough heat to suck and not blow air from the top of the hemispheric aerofoil. I would like to think that in memory of my father you have followed his recommendations and put golf ball like dimples on every conceivable surface that rubs or rolls together.

Golf ball dimples will act as a reservoir for the high temperature grease which must be spread over both sides of the rotor before you literally sandwich it between the two fixed discs.

When you have followed all the health and safety protocols and ramped the speed up to its launch state, all the high temperature grease will simply evaporate, converting all the dimples to cavitating air pockets, which will have the effect of producing hundreds of minute, almost friction free, inter-reacting surfaces.

To launch your naked UFO/plasma active craft, you will require vast amounts of liquid nitrogen, converted to very high pressure jet gas stream. There must be at least four such gas streams escaping from four well aimed fixed external jet nozzles.

The target for the four equally spaced nozzles is the dimple rear ends of the air driven turbo generators. This has two benefits – one, it will spin the hubless high speed flywheel rotor up to a speed where the slip stream driven turbo alternators can provide enough current to continually charge

So You Want To Build the Naked UFO/Plasma Active Craft

the discharging plasma circuit. The nitrogen gas can now be added to the plasma so that it grows in intensity, the few seconds before launch. "Now your craft is plasma active".

Rocket launch pads, are usually in the open and rest on vast concrete slabs which are known as combusted rocket fuel dowsing pits, with all the umbilical feeds falling away at launch. To launch a payload carrying naked UFO/plasma active craft will necessitate the use of a silo. To minimise the launch pad construction costs, it could be possible to find a location in an old quarry where a vertical cut and cover construction could take place. The launched tube, with its substantial dimples, should have an inside diameter of 12.5 metres with a minimum height of at least 77 metres straight up the rock face of the quarry.

Engineers and physicists have written the equations that command the attention of anybody constructing high-speed rotational mass to harness kinetic energy. The greater the efficiency of the mass of kinetic value is based on the highest speed and the distribution of the mass on its shaft. Conversely the bigger the mass the slower the speed of rotation. This simple equation equates to the ability to infinitely scale up your plasma active craft or manufacture it in miniature as if it was a "big boy's toy".

Plasma active craft/naked UFO, call it what you like, but my father's notes were punctuated with Donner and Blitzen, eureka, as his imagination ran away with him – as he postulated the equations that a champagne cork blown out of a magnum of champagne, travelling up a small bore tube, would have X the volume of air to displace. The profile of a champagne cork is conveniently of hemispheric design, with a small lip or collar that allows it to be pressed into the bottle. My father postulated that if the air was passed over the hemisphere of the cork, to add additional thrust to its launch when fired vertically out of the tube, then this would prove the existence of the Coanda effect.

Trial and error are the tools of the concept engineer – equations and hypotheses are the seed corn for scientist. It is only when the two come together in one individual that true genius is born.

I never got to visit my father's laboratory and only on very rare occasions, and then by invitation only, did I venture into his study. Prompted memory recall has helped me to see my father with a thin sheet of copper hitting the corner of his steel rule with a small metal rod we had seen many times before, for my father had a habit of rubbing the rod with a rough cloth to collect the dust. He did not stop when I entered the room until he finished putting indentations over the entire sheet as he rolled it round the bar, slipped the bar and its indented sheet into another tube, then withdrew the bar and peered at me through the tube as if it was a telescope. I put the supper tray down on the table where my mother had told me to place it and stood for a few more minutes as he placed the champagne cork into the end of my pop gun, which I used to fire corks at tin cans. I left the room, as I thought he was cross with me for he uttered the oath again, "Donner and Blitzen, Eureka".

Artist sketch books over the years get misplaced or simply destroyed – it is only by chance that the discerning eye can see their true value. My father's sketch book vividly illustrates his linear thought pattern as he set out to prove the existence of the Coanda effect, the outcome of which would have a profound effect on the hypothesis on the working of its hubless flywheel rotor.

Test equipment was to be simple and cheap to enable the experiments to be funded out of his own pocket and in private

So You Want To Build the Naked UFO/Plasma Active Craft

so not to arouse the curiosity of others. Storyboard sketches were used as a substitute for text.

A champagne cork would cut down to fit in the end of a pop gun. When fired, the cork would travel straight up a short length of dimple tube as an interference fit. As it exits the tube, the hemispheric shape of the champagne cork would snatch a lightweight washer on which were attached four equally spaced fine wires, which would run clear off bobbins without snagging. The process was to be repeated using an undimpled tube with the same interference fit and compare the amount of run off of wire. This experiment, repeated time and time again, would prove with out ambiguity how the dimples improved the vertical travel of the modified champagne cork by a considerable margin.

Size really does matter – for the bigger the plasma active craft the more liquid nitrogen it can carry as its prime fuel. Bigger is also better for all the engineering can be on a rotational and stress level that is easily managed. Bigger does not necessarily mean heavier – this results in considerably increasing the choice of construction materials.

I will not insult your intellect by suggesting a large PAC is shot up a dimpled guide tube by the energy released from a coiled spring. All I have to do is say to you all, just choose between a vertically constructed linear motor drive, a steam catapult, or hydraulic oil used to compress a piston up against nitrogen gas "depending on the size of your PAC you pays your money and takes your choice".

Three score years and ten, I am now passed my allotted span of life and should be indulged as I repeat myself – the naked UFO/plasma active craft is like a radial engine of modular design. Modular indicates it can be infinitely scaled up or down. This scaling process has the speed of the rotor proportional to the size of the PAC, i.e. a dustbin lid diameter rotor will need to be between 25-45,000 RPM – the dome of St Paul's cathedral a mere 1-2,000.

Running up your dustbin lid PAC at just 1,000 RPM should help build the confidence of the launch team to gradually ramp up to 5,000 RPM. Even at this relatively low speed the PAC should be in its dimpled launch tube.

Confidence of the launch team should be built up by continually ticking off all the right boxes. Then and only then should the electric power from the slip stream turbo alternators be discharged across the gap between the two insulated electrodes.

Launch of your unmanned PAC must always be a one-shot event and should be captured by film etc., by using remote controlled mechanical link. No-one should be able to come in contact with wind blown spray as each water molecule/particle will carry a high current, high voltage charge.

Only when the most robust and extreme health and safety measures have been adhered to should the plasma drive be switched on "the continual discharge of electrical short-circuit current between opposite insulated electrodes" for this will launch the naked UFO/plasma active craft and nothing will stop it.

So You Want To Build the Naked UFO/Plasma Active Craft

So You Want to Build the Naked UFO/Plasma Active Craft

Power of Practical Plasma

Plasma, being neither solid, liquid nor gas is the fourth state of matter

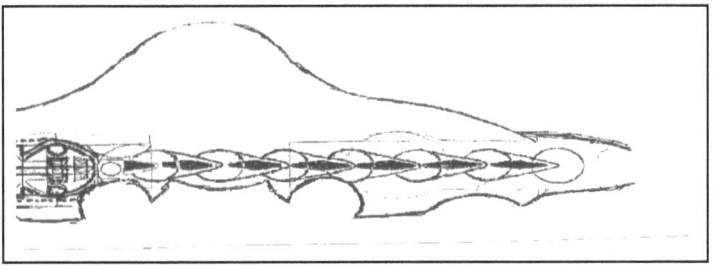

Nature provides vivid lightning displays
As a natural proof of plasmas existence as Donner and Blitzen

Thunder and Lightning was once a scientific curiosity. One day it will be harnessed and put to practical use to drive/power the plasma active craft.

When a high current is passed through an inert gas "Nitrogen", the nitrogen ionises to form a plasma and in so doing pushes the temperature up several thousand degrees. This very high temperature drives the Coanda effect craft.

Set Plasma Free

The Guts of the Plasma Active Craft

Test rig to prove the
concept of
the hubless high speed
rotor/flywheel

Cones to compress,
heat, hold the Donner &
Blitzen and to contain
high temperature ring
of plasma fire

Slip stream driven alternators/homopolar generators to produce the high current for the nitrogen plasma

So You Want To Build the Naked UFO/Plasma Active Craft

Prologue

It was always the same recurring nightmare, the black numbing terror that chilled his blood and shortened his breath till only a desperate scream released him from it. He'd waken soaked with sweat, though thankfully not with the hot rancid stench of urine which had been so prevalent during those desperate childhood years.

The nightmare was real; it had carried on during his waking moments too, until at last he confronted his fears and returned to the scene. He sat in the shadow of the pines that had miraculously survived where so many people had perished, and looked across a dark jumble of still torn-up earth over to where the cold blue water of the Baltic flowed. It had all changed. All the once familiar landmarks obliterated. Here and there a tangle of weeds stood witness to a period when expectations were high, when they could be proud of their achievements in not of the ends they put to.

He recalled his parents, proud father and mother praising him for his prowess in the classroom and on the sports field. Of little Ilse clinging to his hand as he led her carefully across the lawn towards the garden swing. Rolf, the big German shepherd who was devoted to Ilse and followed her everywhere on the Clamstein holiday estate.

He recalled the long talks with his father and the gradual explanation that all was not as nice and safe as it seemed. The revelation that he, Hans von Clamstein und Masselin, had a special duty for his son that might have to be carried out regardless of life or limb.

Time and time again he'd taken Erich to the special hidden safe and shown him how to open it till Erich could do it blindfolded. Time and time again he had shown Erich the special files which in an emergency would have to be transferred to that safe, and if Hans was unable to do it, he, Erich, would have to be entrusted with the duty. It was not to be spoken of. It was a secret so powerful if it fell into the wrong hands it could even put an end to hope that the thousand year Reich would collapse.

But nothing happened to disturb the peaceful well-regulated household until that night of August 17th 1943. Erich knew nothing of the secret trials being held at the other end of the island. Nobody was rash enough to trust a very young boy with secrets of that nature and magnitude, and even his father, a research chemist, had very little idea what was being developed there. No, Hans had a separate agenda, and being a man who utterly despised the new order, had secrets of his own. But as he was leaving the research complex he had a dire premonition, and without considering his own safety, phoned his home and spoke to Erich.

"Erich. Schlaffen mit Gott."

It was the code to take the files to the safe. Erich put the phone down, then went to his father's study, moved the bookcase aside, carefully extracted the files, and left the house via the study window. It was dark but he knew the way thoroughly, and though he stumbled once or twice he kept to the path. It was pitch dark, not even a distant street light to give him his bearings, but suddenly he heard the spine tingling wail of the air-raid siren. He shrugged and kept on. Another practice. There were never air raids in this remote part of the Third Reich.

So You Want To Build the Naked UFO/Plasma Active Craft

But as he approached the low hidden structure which housed the safe he became aware of another sound, a distant heavy throbbing in the sky where hundreds of keen-eyed men were searching for a target in the blackness below.

Searchlights, great white fingers fanned out across the blackness seeking the deadly intruders and failing for the most part. The steady crack of high velocity guns hurling their explosive payloads at the intruders preceded a lethal rain of metal fragments, the razor-edged pieces of good Krupps steel torn apart by the forces built into them. Far away in another country, small boys eagerly collected similar objects and their harried mothers roundly cursed them for bring such dangerous items into the house.

Erich found the safe, opened it by touch, placed the files inside as his father had taught him, closed the safe, locked it, and then suddenly froze as an eerie whistling sound smote his ears. The thunder overhead grew to deafening proportions and then the blackness was lit by long strings of flares descending to the ground. Then it began. The long terrifying and to a small boy utterly incomprehensible earth shaking violence of a major bombing raid. Strings of 500 kilogram bombs walked back and forth over the research base in a non –stop roar of explosions. The very ground shuddered and heaved as buildings were torn apart and years of long patient research were smashed to rubble. People died in the buildings. A bomb didn't care about age or sex or nationality, when it exploded all debts were paid…

For the most part the bombing was lethally accurate, but so too were the defences and forty of the intruders were blasted from the sky. No matter, they were over four hundred strong and the target had to be obliterated at all cost. One of the early victims was a big four-engined bomber with a heavy load of 500 kilogram target indicating bombs. Struck by a 10.5 cm shell in the wing root it fell apart and plunged earthwards. The crew managed to reach their parachutes in time to bale out but light flak, tracing the descent of the bomber, carved through them in an insane butchery of vengeance.

Meanwhile the bomber plunged into a house well outside the target area and exploded. The heat from the target indicators turned the house and everyone in it to a cinder, and to complete the act of retribution, another bomber locked onto the distinctive flaring of the target indicators and released a string of fourteen 500 kilogram high explosive bombs.

Deafened by the noise and terrified out of his wits Erich ran back to the house as best he could, falling several times in the process. Then, as he approached what had once been a neat garden he stopped, and an animal-like howl burst from his throat. The house he had loved was no more, just an inferno of flame, and he knew his mother and sister had taken to the shelter built into it and used mainly for storing things in. As he started to stumble across a dreadful torn-up wasteland the last act was played out. Nobody had ever told poor Erich about delayed action fuses, and the blast picked him like a straw, ripping his clothes off and hurling him back into the trees….

Erich's memories of the next few years were mercifully dim. Like so many other children in Germany he should have died and this story would not have been written.

Strangely enough it was the enemy who succoured him. Caught up in a movement of refugees he was still in a state of shock, cold, hungry, wet, thirsty, and near the end of his strength. He wasn't aware of stumbling into what seemed to be an empty road, and the driver of the oncoming Comet tank gave an incredulous look at the figure walking straight at him then braked to a halt.

So You Want To Build the Naked UFO/Plasma Active Craft

"What's up, Driver?" the tank commander asked, looking up from the radio set.

"Bloody nipper right in the way sir. Jesus will you look at the little sod, he's leaning on the flaming track!"

"Well get him off dammit, we're holding up the convoy!"

"Yessir." The driver jumped down picked the boy up and felt, not for the first time, an awful wave of guilt. Warring against the bloody Nazis was one thing but kids like this, no, it wasn't his fault. He looked around for somewhere to put the kid and glory be a jeep flying a Red Cross flag roared up. He stepped into its path and walked across to the driver.

"Ere mate, this little sod dunno if 'es comin' or goin', walked right into us 'e did. Can you see to him, we've gotta get ter Grunevild by termorra and we ain't set up fer kids."

"Okay corporal," a female voice said from the rear of the jeep. "We'll see to him for you."

He saw the flash of epaulets and knew the kid would get attention.

"Thanks ma'am," he passed the boy in, saluted, then ran back to his tank, and in minutes the convoy was once again a roaring column of steel heading into the wet gloomy winter afternoon.…

Erich never realised how lucky he was. Not only was he passed into trained medical hands, the officer seated next to the nurse was in the British Intelligence Service, and his ears pricked up when he heard the feverish muttering of 'Peenemunde'.

"Hel-lo! We may have something here, Noreen."

"He's in a bad way John. I don't know if they'll take him."

"Oh but they will! He just said Peenemunde."

"Is that significant?"

"Maybe. That was the research base where the Nazis made the doodlebugs and V2's."

"He's too young to have had any part in that."

"True, but he may have family and I believe the powers that be are interested in any of the survivors. Let's get him to hospital, I'll sign him in."

And so forty-eight hours later a hospital flight carrying, among others, Erich von Clamstein, touched down in a remote airfield in England and the boy was rushed off to a US military hospital…and from there to America.

It took a fair while to nurse the boy back to health, and his plaintive cries held enough to let know his parents had worked on the research base. Eventually he became lucid enough to let them know his full name, and the German speaking nurse felt a flash of triumph when he said 'Erich Clamstein'.

"Clamstein? Nope, new one on me, nobody of that name on rocket research that I know of," Colonel Fox, the big bluff intelligence officer attached to the unit said. "Hell of a lot of records missing though, the Brits really pulverised the place. Just as well, if they hadn't Hitler may have stopped the invasion. I'll have a word with von Braun, maybe he can help identify the kid."

So You Want To Build the Naked UFO/Plasma Active Craft

Werner von Braun, the chief research specialist of the American rocket program, frowned thoughtfully when Colonel Fox asked if he knew of anyone called Clamstein.

"Ja. There was one. I knew him to speak to. Hans von Clamstein und Masselin. He was a chemist, a good one. Worked on fuels, but he had some crazy idea about using the water, nein, steam, to power rockets. Something like that. Was killed in the bombing, and all his family. He had a wife und…and two kinder, children. A bomber crashed on the house and blew up."

"Ah. Interesting, doctor. We have a young boy in the hospital, the Brits found him as a refugee and he was in a real mess, but an Intel officer heard him saying Peenemunde over and over in a fever. Contacted us and we flew him here for treatment. About eight, he could be older I suppose…Said his name was Erich Clamstein."

"Ja! That was the name of Hans son. Everyone believed he was killed in the fire but he may have got out in time. He had a sister, her name was Else..nien…Ilse."

"Would you be able to identify him, sir?"

"No. I never knew Clamstein well, we met at conferences you understand but he was in another section, department, and the Nazis did not like us to fraternise. The Gestapo or the SD watched us all the time. I met them in Peenemunde in the village once or twice but not socially."

"Clamstein was killed?"

"Ja. He tried to get home when the bombing started. His car was found in a bomb crater upside down and he was still in it, kaput. Broken neck and spine. I remember thinking he had nobody left to mourn him at the funeral."

"Oh well, it was a long shot anyway. I'll see that the kid is healed up then we'll send him back to Germany. There's an organisation out there now that looks after victims like him and he'll likely be happier where everyone speaks his language."

Time passed…In the summer of 1956 Eric Clamstein, citizen of the Federal Republic of Germany, age 18 years one month, declared unsuitable for recruitment into the German Armed Forces, made his way to the Baltic coastline. International tension was low and nobody paid much attention to the thin awkward-moving youth with the slurred speech. He managed to cross onto the Isle of Use Dom and once there, made his way across it to where the Clamstein estate once stood. Bitter tears stung his eyes as he looked on the ruins of his childhood. Then, with a muffled sob he turned his back on the scene and followed the directions drummed into his mind all those years ago.

It took time. Nothing was as it had been, but eventually he reached the site where his father had built the small shelter.

It was overgrown with brambles now, but Eric determinedly ripped them down and there facing him was the metal door, now rusted to a colour which matched the surrounding earth. He scraped around, found the box that held the key and yes it was still undamaged thanks to the layers of grease in the box. He took a can of cheap lubricating oil from his pack, moved the cover to the keyhole aside, and squirted oil into it. More oil went on the hinges, then he waited.

The key turned in the lock and he pulled on the door. The bottom edge stuck fast on debris which had gathered over the years, but he kicked and scraped at it till the door was fully open. There, nestling in the metre-wide tube, was the safe. Eric applied more oil to the tumblers and set the number. To his relief the lock clicked open and when he opened the safe he saw that everything was exactly as he'd left it.

Still obeying the imperative drummed into him by his father, Eric took out the leather bag and looked inside. A blaze of colour met his eyes as he looked on the von Clamstein fortune. Flawless gemstones. Diamonds, rubies, emeralds, only the best stones available had been good enough for Hans, and he had chosen well. There was a letter, a voice from the past telling Eric to go to Switzerland and contact a banker. He was to say he was from Hans Von Clamstien und Massellin, and everything would be taken care of from then on. There were also the papers from his father's study. He packed everything away, then shut the door and made his way off the Island. Five weeks later the GDR made the island off-limit to all civilians, but it was too late. All unknowing, they had shut the stable door after the horse had bolted…

Naked U.F.O

Introduction

Naked U.F.O is the vehicle that lets Erich Clamstein vent his frustrations on the lack of order and direction of modern society in a humorous and, some would say, funny way - if the subject matter under discussion was not so serious.

Eric Clamstein systematically sets out the problems and then proceeds to ram the solutions down the reader's throat, until in the end his obvious frustrations are those felt by the founding fathers as they set sail for the Americas to set up their new society.

Eric Clamstein's solution is to set plasma free so that we might re-colonise Mars, if not fly to the stars.

You could say that Eric Clamstein is a court jester to the modern global technological campus, as through the pages of Set Plasma Free he systematically tells the reader how to build and fly a Hollywood interpretation of an alien craft, a euphemism for a UFO.

So You Want To Build the Naked UFO/Plasma Active Craft

Naked U.F.O

From DNA to Rockets and Flying Saucers

DNA is how mankind has dominion over all the lesser animals. To this end he woke up one morning and said "this ere earth ain't big enough for two types of manlike creature" and promptly kills off the lesser successful model. And he has been on a killing spree ever since. And finally as mad was writ large as mutually assured destruction, for each side had an arsenal of rockets which in turn could spread multiple warheads as if they were pollinating the air.

Mad would have sent us all back to the Stone Age with its nuclear winter. Instead we keep the home fires burning 'til they are extinguished by the rising sea levels caused by global warming. Oh well, win some lose some.

So the Russians use their rockets to help the Americans build an island in the sun (International Space Station) and the Americans strap their would-be space tourists into a space plane whiched hitch a lift on a bomb (my apologies for lack of sensitivity).

If Hollywood can have a UFO in cult films like "The Forbidden Planet" and "Invaders From Mars", why shouldn't I, as a charlatan parroting scientific gobbledygook, blow the lid off conspiracy theory after conspiracy theory and tell you how to fill your supermarket trolley with all the technological goodies you need to go to Mars, if not the stars.

Engineers, engineers everywhere and not one left to think

It is silly of me to look back to the Industrial Age of the 'lunatics', for they were the enlightened ones, the brave men, wise men who lay down their philosophy upon a hill by moonlight.

The question asked - were they moonlighting? No, they were the engineers who built the industrial revolution. From their vision of science and engineering we built a sophisticated killing machine, capable of harnessing the power of the atom to do our destructive bidding.

Society, by its very nature, must always dissect the present to discover how we got here from the past. Slavery, the opium trade, the industrial revolution, the killing fields of the Somme, Passchendaele, the Treaty of Versailles, The Great Depression, the meteoric rise of Fascism to the birth of the Second World War.

Reconstruction of the global economy led to the removal of any restraints that had shackled the free spirit of the world's youth.

Teachers follow this new false doctrine of a liberalisation, bent the rules to embrace the gospel of self-expression, turning a blind eye to lapse in discipline under the guise of discovering their roots.

To metal detectors writ large to discourage primary school children and their older peers from taking knives to school.

With stop and search watered down to a dissertation on their human rights and parliamentary statements that you carry a knife, banged go your rights, locked up for years and early nights.

So You Want To Build the Naked UFO/Plasma Active Craft

So how much further must we go to spare the rod and spoil the child should not be placed in a glass case at the behest of the nameless ones in Brussels?

Picture a future scene, a police people carrier disgorges its occupants from every sliding door onto a suspect.

Not to manhandle him but to encircle him so that one of their kind can spend personal time with him as he goes through his civil rights.

One of the officers tickles him and finds a six inch lock blade knife and his passed around his colleagues. Some officers are heard to say "Hi son, you've done a good job of sharpening the edge and the blade on your knife is very sharp, not to mention how well you've polished it. Look I can see my face in the blade.

I have to caution you under health and Safety, not to touch the blade; I also have to issue you with a request that you attend a Police Course on how to use the handle of your knife and not the blade to hold it by.

One other thing, Son, put it away or you will dazzle a motorist and we will have to book the driver for careless driving.

Hold it son, not so fast, we must all congratulate you on the pivotal role that you and your associates are playing in keeping us all in jobs and the £60 billion a year crime rollercoaster in motion"

Opium could never compete with gin and lost the battle with the gin palaces. As global alcoholism became the natural sanctuary for those condemned to waste their pathetic lives, trapped in a labyrinth of alcoholic haze, driven by the poverty of ignorance.

Set Plasma Free

Evolution over just a few generations proved that an appreciation of the power of learning would give them a life line out of this 'malaise' and domination over those left behind.
Ironically Passchendaele, the Somme, gave way to "Oh What a Lovely War". Rats made a good living out of those dying on foreign fields; spread the plague of the Great Depression and financial ruin.

London's Burning; St Paul's is singed in the second great fire of London. Hamburg is lit up in an incinerating wave; Hiroshima and Nagasaki are destroyed in a blinding flash of Pandora's Box. A war is ended and no-one lost apart from the souls of 50 million who gave up their lives for a quantum leap in consumerism.

Do I see the endless parade of the guilty ones? Who advocated this global liberalisation? Let them now stoop down and pick up the pieces of shattered young minds, for under their stewardship they lay apathetic, dysfunctional, undisciplined and will crash at the slightest challenge to their free-wheeling souls.

Sons and daughters passed into others hands without the pain of rules laid down and rules enforced, when the lease is broken and the spirit free will self destruct in a negative force as they kick against authority.

Cocktail sticks and champagne glasses, public money piled high, built an architects vision of villages in the sky. Politicians' indecision between hard and soft stumbled zombies with all their vision lost.

No man's land where drugs change hands, a lost generation from this promised land.

So You Want To Build the Naked UFO/Plasma Active Craft

Past generations had brains that could be trained to read and write, add up, subtract, divide and multiply, root learning raised the productivity of the industrial revolution. Led conscript soldiers to kill or be killed by route.

Liberalism had a vested interest in conveniently forgetting that a good basic education had built and sustained the British Empire with all its engineering needs.

And so today, what do we all have to shout about? The finest engineering excellence taught behind hallowed walls – this exclusive club is not for our nieces, nephews, sons and daughters, with multi-choice questions just a tick in the box, and exams dumbed down. Others of a difference hue will wear exclusive crown. The tiger economists and land of the rising sun, great wall of china what have we done. With workplace assessment and access for all, health and safety less we should slip and fall and sue in the courts, with all these expenses and burdens piled high, the computer switched off less it should strobe your eyes, the keyboard you use no longer complies – must be recycled at once, your workstation condemned, you are either too short or tall, the seat that you sit in you are either too fat or too thin, you've got a degree and debts piled high, big brother is beckoning with Mickey mouse job, there is a sting in the tail with VAT and PAYE, one third on return is 50% on cost, once in his clutches your ambition lost.

The rough and tumble of militaristic playground games, to lessons taught with brilliant flashes, thunderous roar and shattered drums, their cut their fingers and burn their thumbs. English is taught as the first language of science, with all this behind them the papers in English no problems at all.

You must build a prototype the call goes out from the venture capitalists, the VCs. If you build a device and we like what it does you must write a business plan, cash flow forecast and all.

Now you asked for some money, that's not funny at all, but if you use all your hard-earned cash, legacies and all, we will put a proposition to you that you cannot refuse. That A must be B and C as well, with A shares to us and B to you, C is off-shore if you see what we mean, we will pay you some money to work day and night, to meet a deadline and get the product right. Your cash flow said you would sell it to Tom Dick and Harry, we don't like Tom or Dick and as for Harry……! So you are now in default and other backers brought in, they shuffle the cards and deal you out, they grab the pot and it's goodbye to you.

Do I sound bitter? Not at all, the cash for nuts and bolts is better than dinner with friends – at the end of the day it is a lottery ticket. If you get it right it is money for all.

So You Want To Build the Naked UFO/Plasma Active Craft

Fusion, A Crash Course For All

I want you all to grit your teeth for a course in atmospheric plasma (AP). I found out about AB on the ubiquitous golf course. Nothing unusual in that, I hear you say. A late afternoon in a hot, sultry summer day; my golf was so bad that I had to play alone. My battery driven caddy had witnessed my folly on an earlier date, now refused to move. I am playing a shot to the 18th hole – my bladder now bursting, time for a comfort break. A prayer to the gods that it should be struck down, a blinding flash of a lightning bolt. The caddy's prayer is answered, it is vaporised in a flash, and the clubs now follow like a roman candle.

I know nothing about golf, but 30 years of AP has sharpened my mind. What is AP, I hear you say. Now I will give you a copyright verbatim reply from the E.F.D.A. (European Fusion Development Agreement). Their PR handout titled "The Joint European Torus – A European Success Story. Fusion, the energy of the sun".

If the temperature of the gas rises above 10,000 degrees C "that's not North Sea Gas at Regular 6, but the price hike that hits you could well push your blood pressure that high".

'Virtually all atoms, silly bits of matter that hold the whole universe together including you and I, become ionised (just a throw away line to describe the jumping of a spark from one point to another – the spark is the ionised air). See it as a game of musical chairs – the empty chair is an atom that has lost one of its electrons and now it is called a positive ion. The minute it gets an additional electron we call it a negative ion, but this only happens if it is bullied by a powerful electrical current – other than that it is not an ion at all.

Set Plasma Free

Electrons buzz around their atoms like bees around a hive. In our case, the hive is the centre of the atom which we call the nuclei (that's where we get the expression, nuclear family i.e. the children (neutrinos) are held in check by the bond of their parents but if the pressures from outside from working too hard to drinking too long, the money they make is gambled away as they split apart under tremendous pressure, the kids (neutrinos), which are the glue that hold them together, fly off the rails – the nuclear family with all its energy lost.

So who are Mr and Mrs Nuclei? He is a positron and she is a neutron, but they hold their kids (neutrinos) tightly so you cannot see them at all until you dump all the power from your local power station as if it was plugged in the wall.

Energy in is energy out – the result of this complex mix of neutrinos (kids out of control) and ions (the extended family trying to rein them in) with the sum of all charges being very close to zero (no win, no lose) as only a small change of imbalance is allowed (we have to compromise to survive) thus the ionised gas (all that very hot air) remains almost neutral (does not take sides) throughout.

This constitutes the fourth state of matter called Plasma (our old friend atmospheric plasma – lightning bolt).

To strengthen the DNA of our nuclear family we have to (fuse) marry off Ms Tritium to hansom energetic Deuterium in a ceremony we call a fusion.

This can only be done with very special energetic plasmas on their special day, to keep all those invited in and all others out.

We use a magnetic consignment designed as a tokomaks. The joint European Torus is a tokomak in which strong magnetic fields are used to confine the plasma. In this class of device

So You Want To Build the Naked UFO/Plasma Active Craft

the plasma chamber is a donut shape to avoid end losses which would occur, for example, in a linear cylindrical configuration.

The vessel is filled with gas at a very low pressure and this gas is converted to hot plasma by passing an electric current through it (our old friend the local power station plugged straight in to it).

The question you will ask, if it takes all that power to get Ms Tritium and handsome energetic Deuterium to consummate their marriage (fusion), how big is the whole kit and caboodle (J.E.T.)?

How about removing the dome of St. Paul's and dropping in J.E.T. and replacing the dome as if it was the lid on a pot – be careful if doesn't boil over. Don't worry I am only joking.

Flying Bedsteads to UFOs

So what is new in 1066? The wind still howls over barren land and the sun still hides its blistered hand.

Oops – not 1066 but 1956 when some boffins (a symbiotic cross between a scientist and an engineer) came up with the idea that they should design a new all-singing all-dancing jet fighter that would literally dance in the air by using a jet engine to push it up vertically so that it might hover like a helicopter.

Some would say, why not go the whole hog and build a flying saucer as per Hollywood? . So there it sits, two conjoined Rolls Royce jet engines held in a frame which looks remarkably similar to an old-fashioned bedstead. That dot on the top is a man who is going to fly it – you must be joking! If you think that is funny "Rodders" wait until I tell you we are going to build a "flying saucer" and this time next year "we will all be millionaires!" (Acknowledgement to the catch phrase from Only Fools and Horses). Now I have said it, yes from here on in, in the next cascade of pages, I shall tell you how to build and fly your own equivalent to the Rolls Royce flying bedstead, which I shall name T.R.I.P; a test rig to prove the concept of triangular rotating interrupted plasmas. Or put it another way, jet thrust driven homopoleic excited atmospheric plasma drive for a circular aerofoil (a flying saucer to you).

The only reason that I digressed into fusion of the nuclei of deuterium and tritium was to vividly illustrate the stratospherically amount of power/energy needed at the front end to get a quantum return of energy at the back end. Try this for size, "the fusion of all the atoms in just one litre of deuterium-tritium gas mixture would create roughly enough energy to provide the yearly needs of the average sized

So You Want To Build the Naked UFO/Plasma Active Craft

house" (quoted verbatim from the dust jacket of a computer CD aptly titled "Fusion an energy option for the future, an EFDA copyright"

We will now start in earnest our campaign to "set plasma free". Who said that all plasmas, regardless of energy in to energy out, had to be held in an unimaginably strong electromagnetic field and whilst suspended in such a field they could not break out and vaporise the massive steel doughnut ring which allowed the electromagnetic generated field to do its work? Only when we can demonstrate to all the sceptics of the scientific and engineering community that plasma is much happier playing around the circumference of our circular aerofoil, or even slipping to the underside of it.

Why would our plasma like to play in the freedom of the earth's atmosphere? As with all my questions this one is purely rhetorical; put simply, outside of any containment vessel it can gorge itself on the very plankton of physics, namely the humble electron, which it would require by the trillions in an endless feeding frenzy as it plays musical chairs at the speed of light. So far it is only words – how shall we do it? Let's take a trip on triangular rotating interrupted plasmas or, as Hollywood would have us believe, ride a tornado to the Land of Oz. Shall we let our test rig, like Dorothy's Kansas house, fall on the witches and wizards of conventional engineering and scientific practice, for all good stories are sword and sorcery (OK so it is an obvious pun), there has to be an energy source for good and evil.

Let me now introduce a little-known wizard of the scientific and engineering community – a Romanian engineer Henry Marie Coanda, for it was he who found out in the 1930s that a stream of air blown over a curved surface will have the effect of the air stream sticking to it, thus making the air pressure under the underside of the flying saucer lighter. It is this pressure differential that gives the saucer its lift.

Is it a form of hovercraft, I hear you ask. If only it was that simple. Wind in your hair, wind in your sails, the wind of change and the wind speed at the core of a spinning tornado can exceed 300 mph which would easily lift Dorothy's house. The amount of energy to drive a super cell weather system would take the output from numerous 500 megawatt power stations. If our flying saucer is ever going to suck itself into the air by harnessing the coanda effect it must be constructed by using some of the lightest and, conversely, strongest material known to Man. How about balsa wood and expanded polystyrene (Oh yes, and a couple of cardboard tubes from the centre of run out toilet rolls, not to mention papier-mâché made from miles of toilet paper – go on this has got to be a Blue Peter wind up!) So now I am a mind reader. I will stay with the expanded polystyrene but add ten if not 100s of miles of 5 micron drawn glass fibre filaments and enough polyester tape to fill a mini container, not to mention rolls of carbon fibre wrap which would easily cover a soccer pitch.

What part could a good old London step-on step-off, any more fares please route master bus play in the great scheme of things?

The funny thing about buses, not one turns up according to the timetable and when you have given up all hope, four turn up at once.

My convoy of four buses has just driven onto the sacred turf of Lords cricket ground and made a very appropriate red cross of St George. As each bus is at least 30 foot long and fourteen feet high, it could well represent a tight fielding configuration for a cricket match.

Ok now drive the buses off the cricket pitch before we all get arrested for vandalism and have to do 200 hours community service watching England get slaughtered at our own game.

So You Want To Build the Naked UFO/Plasma Active Craft

You can see by the way how our heavy route master buses have cut up the field and in so doing, left a visual impression with their heavy tyres just what a 30 foot diameter flying saucer would look like – not to mention the awe inspiring 14 feet in height is enough to blow your mind away.

Now to build it, Hollywood fashion. Peg out a 40 foot square and then run some string, approximately 1 metre apart so you have 40 one metre rows. In to this lay strips of 40 metres in length by 1 metre in width of robust heat shrinkable pallet wrap. (You know the kind of material I mean, it's called Big Boys Clingfilm and it holds all sorts of things tightly onto shippable pallets).

What we do next is to set out what we will call a 10 foot shrinkable border. Once this is defined, now lay this time 30 metres in length by 1 metre in width a wonder material that very fortuitously has just come onto the market which is called Curon roofing material. (You know how much I like a bad pun, could the CU in Curon stand for Cure/ harden and could the ron be sun – there you have it….cured by sunshine – which of course, as we all know, is in short supply so let's take the manufacturers advice and let it cure itself over two days by UV light, ultra violet radiation to you)

Remember to read the instructions on each roll of Curon which states, "first roll out the roll into strips", which in our application will be one fully unwound roll of 10 metres which is slightly over 30 feet in length. Remove the backing film which exposes a strong adhesive which will naturally stick to our pallet rack. Now proceed to unroll the other rolls so that each roll overlaps the previous one by 2 inches. The manufacturer states that it takes 2 days for the ultra violet radiation to harden the material on to its substrate.

Now the two days has passed, we will apply a bog standard expanded polystyrene glue over small sections of it so that we can then drop on the top high density polystyrene slabs which are conveniently half a metre thick by one metre wide and two metres in length.

By high density closed cell polystyrene we mean the kind that doesn't break into snowflake dust, as each block is laid down onto the adhesive it is pressed firmly up against the preceding block. This process is continued until we have laid one layer making a 30 foot square. Back to the Curon, peel off the backing and stick it down on top of the polystyrene and remember each layer takes 2 days to harden, so allowing for 28 layers, in about three months time we will have a solid chunk of high density expanded polystyrene reinforced with layers of Curon. To finish it off, repeat the same process of applying Curon to the sides and to the top surface, wait the two days and hey presto! A solid slab sits there, looking like something out of a science fiction feature, reflecting the light with a strange translucent eerie glow.

Now we have to ruin it all by machining it with 5 axis high speed milling spindles.

By the time the computers have finished it will sit there, in all its glory as a 30 foot diameter flying saucer just awaiting a final coat of Curon over the now exposed profile.

I was inspired to instruct you to do this by slipping back five thousand years as if I was the project manager responsible for selecting a suitable outcrop of rock which my stonemasons, with their bronze chisels and wooden mallets would carve into one of the wonders of ancient Egypt, namely the sphinx.

The question which has always gone unanswered, who inspired the Egyptians to build the pyramids? Could it have been that their own legends told them of a time, way back before history, that a race of giants descended upon a sunken valley with a small river running through it in a dry barren

So You Want To Build the Naked UFO/Plasma Active Craft

land. And how these giants through their engineering skills build huge pyramids to hold precious water which would be chilled in the hollow pyramids and then allowed to flow by gravity through terracotta porous pipes so that the chilled pipes would inter-react with the hot organic matter to produce droplets of condensation on the outside of the pipes and so nurture their crops from beneath the ground, giving them bountiful food in a Garden of Eden without any rain.

Now I have said it – no rain. So where did all the precipitous over what we would call Northern Europe fall? It fell as snow to reinforce the glaciers and maintain the ice age.

And how did the giants build these pyramids? Did they carry huge chunks of stone around with them and lay them down as if they were polystyrene slabs, layer upon layer and then carve out the inside to make it hollow?

No, no they were far more practical then that, they took the reeds from beside the river and wove them together to form what we would call a small sheep pen, then they filled the pen full of sand and gradually they built up a vast pyramid with a hollow core and as each layer went up, just as it takes 2 days to cure our Curon, the whole pyramid settled uniformly as layer after layer of sand was compressed into a solid block. So there you have it, history is now repeating itself, from the great floods of religious history across all faiths as the ice age came to an end and the glaciers melted sending tidal waves of water that flooded the Garden of Eden and gave us the Mediterranean as we more or less know it today. And yes, Noah was told by the giants that they were going to release this water, they were going to release the dam and told him and his disciples to build an ark – our 30 foot saucer will hardly compete with Noah's Ark, but surely it is a step in the right direction as through the next pages I will tell you how to fit it out and how to fly it.

Inspiration for my UFO came from a moment of madness at a hatter's tea party

Did I go mad at a Hatter's tea party? Why was I at such an event? Something to do with my great Aunt who had built up a formidable reputation as a Court dressmaker and had dragged her long-suffering nephew along as a buttress against any unwanted attention from male or female alike.

Millinery and hattery were now in the ascendance, having shrugged off the frugality of the war years. All were carried away with emotion at the first such event since the cessation of hostilities. Creations for Cheltenham, Ascot, weddings and funerals were displayed by young ladies. The men, not to be outdone, stood in splendour bedecked with hats for huntin', shootin' and fishin' – and a night out at the opera.

This left me feeling sick from my over-indulgence in comfort eating, as I munched my way through the spread which heralded the start of the proceedings.

I excused myself on the grounds of feigning stomach pains and returned to the scene of my crime, to pick my way through the remnants of what had been a buffet fit for the sport of kings.

The remaining grapes were quickly scooped up and crammed into my greedy mouth. Then I stopped in my tracks as I contemplated the vastness of the now empty fruit bowl, to call it a bowl is to give lie to its size, for it sat there tempting me to pick it up and hold it up to the light, to see if it had the true hallmark of the finest bone china. The acid test, could I see the shadow of my hand when held up to the light. Once this experiment had been proven, a glance at the manufacturer's trademark told me to put it down. For even with the little knowledge that I had picked up from others, screamed it was

So You Want To Build the Naked UFO/Plasma Active Craft

worth a King's ransom. Did I put it down with the reverence it deserved? No, not I. It was duly balanced upside down on three lipstick stained teacups of a similar "provenance" – all three of which were set out in an equally spaced triangular configuration.

What was to be the crowning glory of my doodling with all this priceless Meissen china - yet another teacup, this time unsoiled, so it could be placed upside down over the manufacturer's dead centre trademark.

As if by pre-arranged signal, no sooner had I stepped back to admire this translucent edifice rising above the untouched oranges, apples, pears and bananas that seemed to pay homage to it, then without warning one, two, three flashbulbs went off in quick succession as a bored paparazzi flooded my UFO with flashlight. Caught on film, one young man who had too much imagination for his own good - alas all prints and negatives are lost in the annals of time. I do now remember that the photographer sold his pictures, which when published had the misfortune to upstage the whole event.

Publicity generated from my misdemeanour brought my Great Aunt enquiries from attention seeking bright young things, who would have commissioned my creation for Royal Ascot.

Set Plasma Free has been an albatross around my neck and as such has cost me dear. For my Great Aunt was fickle in her patronage of the offspring from her married brothers and sisters and duly disowned me! Let no-one have any doubt it was that creation that brought back those horrible dreams, that passed from nightmares to reality and all the time the blueprints and the pencil sketches of the coanda effect haunted me.

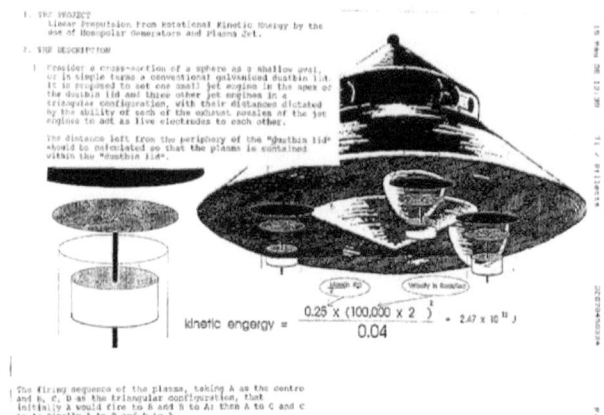

It once was a slab of polystyrene covered in a translucent rock-like material doing nothing at all, just a giant square obelisk.

Not anymore – after machining it stands on its three machined teacup legs, like a ballet dancer in a tutu, about to perform a role from the ballet Swan Lake and our slab has turned from the ugly duckling to the profile of a graceful swan.

So You Want To Build the Naked UFO/Plasma Active Craft

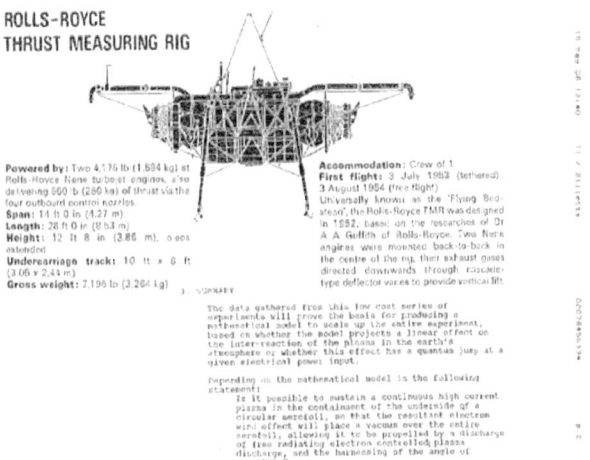

The military will always seize on an opportunity to usurp technology to its own end. Set Plasma Free has not been immune from such attention. With not a straight edge in sight, it is a stealthy design dream for all those who desire immortality for their design of military kit for land, sea and air, as they will go green with envy. However, their computers would crash with the very thought of designing all the control surfaces for a true coanda effect flying saucer.

Before I tell you how it is done we must put back all the Curon roofing material that we have machined off. Remember, from this point on each roll of Curon weighs in at a hefty 44 lbs – or should you prefer, a trip to the supermarket for 22 x 2lb bags of sugar. If we have used 100s of rolls so far, we could well pass the 1000 rolls mark before we are finished. How do we use a 1mm thick, rigid material to flow effortlessly over continuous curved surfaces?

Not a problem. Refer to the manufacturer's instructions then practice on an orange with a sharp knife to score the peel, as if to cut the orange into equal segments. Next, peel off the segmented peel only – lay it on a plate and look at it's cut out shape. Then proceed to do the same on an appropriate scale with the rolled out Curon roofing material.

The manufacturer states "by working the cut-outs with the standard two inch overlap into a patchwork quilt configuration they will cure over any curved surface".

On our military theme, some will argue that Curon is a substitute for carbon fibre. Unlike carbon fibre, Curon's density is much higher so you have a considerable weight penalty. If this penalty handicap can be accounted for in the design, why not go the full nine yards and blow the bank roll on additional rolls of Kevlar until you have covered the entire craft so that it is impervious to all that scabbard fired high velocity depleted uranium solid darts, as all other projectiles will simply ricochet off all the curved impervious surfaces.

Talking of impervious to projectiles brings me to my next piece of "pantomime". The villains of the piece are 1920s rent-a-mob. Gangsters of the puritan prohibition era of the 1920s and 1930s, when only the Americans, through their arrogance, could unilaterally set a global example and deny to the peasantry the dubious pleasure of drowning their sorrows in the panacea of alcoholic haze as the Great Depression took hold.

Listen to the sound of the machinery in one of the few growth industries, as shotgun cartridge after shotgun cartridge is filled to supply the ever hungry ubiquitous pump action shotguns, which are used to protect turf from others who wish to take over the lucrative supply of bootlegged liquor to the sprawling chains of speakeasy joints.

So You Want To Build the Naked UFO/Plasma Active Craft

In our little story we are replacing the lead shot with minute pellets of copper. The word paint is a euphemism for spraying with bullets and denotes a concentrated volley of shot.

That's what our rent-a-mob are doing with our translucent saucer. Not one part of it must be left unscathed without minute particles of flattened copper imbedded in the Curon.

Next let them reload with different cartridges, this time the shot is fine white silicon sand.

In the real world the machine spindle would be replaced by what is paradoxically known as hot plasma, cold metal depositaries onto a metal surface. This is achieved by a cousin of Set Plasma Free, where a high current electrical discharge has injected into it micron particles of the material that will explode into a high speed vapour and, on impact with the material to be coated, will plasticised and adhere to the matrix of that metal.

At time of going to print no-one has ever had the desire to coat non-metallic surfaces with this process. All this is about to change with the publication of this book – for like all my snippets of scientific transgression I have a further trick up my sleeve.

A Dazzling Light to the Heat of the Sun

There is no such thing in the world of inventing as a totally new concept. As the chapter heading suggests, a dazzling light to the heat of the sun was conjured up by a Greek wizard whose name escapes me, when too many ships to count decided to attack his land and call it their own. With his meagre resources he applied lateral thinking and polished up hundreds of bronze shields, setting them in a concave configuration so that the sun would strike the centre and the focussed bright light was so powerful as to burn the sails of the invading army. Not to mention melting the tar that sealed the timbers.

Do we from that get the expression "shiver me timbers"? Well in the case of our UFO we have to apply a modern day version of what I have described. In simple terms we replace the plasma head with an exceedingly powerful Zeon bulb, the kind of bulb normally associated with high powered cinema projectors – or good old fashioned search lights for film premieres. Because the light is so bright and the potential infra-red radiation/energy is so powerful we have to use a short focal lens which must have all the hydroxide removed from it "hydroxide is a fancy name for impurities that would vibrate to infra-red radiation thus causing the lens to heat up and become unstable and crack".

Now that we have focussed all this infra-red energy we can classify it as a non-coherent light source - the exact opposite of a laser.

Our light source will vibrate the atoms immediately above the silicate and the copper to about 2000 degrees Centigrade. This will have the effect of the silicon structure accepting the molecules of copper. At the same time that this is happening, once the light/infra-red energy source moves on the effect is

So You Want To Build the Naked UFO/Plasma Active Craft

to leave a one or two micron glazed finish over the top. By spending three months very slowly sweeping the entire UFO we have now produced a paradox of electronics whereby we have blended the molecules of an insulator with the molecules of a conductor. This is only going to be relevant at exceedingly high voltages.

To give you a for instance, if our Set Plasma Free round the circumference and under the rim of our UFO becomes exceedingly hungry, then it has a ready source of electrons released from the copper atoms, but at the same time vast numbers of electrons are snatched, kicking and screaming, from the glazed silicon. We have in point of fact produced, at very high voltage, a magnitude of cascading discharging capacitors.

This of course will now achieve the object we are looking for, and that is to create an unbelievably powerful electron wind to drive the coanda effect and simultaneously to feed our Set Plasma Free.

I must catch my breath after all of that and come down to earth and tell you wonderful people how we are now going to start hollowing out our UFO, if I can work out how to cut our way into it! Before I do so, I just want to reiterate on a sub-heading.

An electron wind – now there's a funny thing!

What is an electron wind? As previously stated, it is a very high voltage and subsequently a low current electrical discharge. For example, in the early days of high powered laser cutting, in order to stop the edge of the material from spluttering (nasty metallic or carbon fibre blisters) the laser became a positive electrode to a negative earth and very high voltages were discharged and the electron wind effect cooled the immediate area being cut by the laser.

Now let us return to that wonderful word, magnitude. I always use this when I want a very big number with huge numbers of noughts, when I haven't got enough fingers and thumbs to quantify it.

As an aside, lightning conductors are normally made of copper to allow the electrons to quickly pass the lightning bolt to earth, thus saving the fabric of the building from exploding. You can therefore imagine that our craft, with its fancy coating, is going to act like trillions of lightning conductors – the only difference being, there is no discharge point so all electrons, like lemmings, must plunge over the edge of our UFO to be gobbled up by the plasma.

Hold on to your hat – I seem to have jumped a bit! The question that the scientists and engineers, even those who like a laugh and a giggle, will be asking themselves "where is all this power going to come from?"

It was once said that it was always easier in life to solve a problem, and as you do so create a new set of problems.

!!So it was with I!! Yes, over thirty years ago I had worked my way through the United States bureaucracy to achieve a private presentation of my work on the electro-mechanical

So You Want To Build the Naked UFO/Plasma Active Craft

pulse reaction drive, only to be told by the then councillor for electronic advancement that even if he understood my blueprints that I laid out before him and the scientific concept that supported it, and if they built it they would only have to give a nod and a wink to their counterparts in the Soviet industry as they would put it to equalise the balance of fear.

As Concorde had been rolled out a few years earlier, in 1969 and Koncordski had followed suit, his point was well made.

With our modern global family, Western governments follow each other like sheep and squander the people's taxes on systems that do not work but do work for all the wrong but right reasons.

How often have you heard wishy-washy liberals, soggy socialists and goody-goodies say that prisons do not work? But to the masses who walk the streets, they say they do work for it keeps the villains and the delinquents out of their hair all the time that they are locked up "over there".

Because the number crunchers only look at one side of the equation, out come the villains on parole, where they are duly tagged, a probation officer is assigned and the whole aftercare swings into operation.

Some bright spark who understands electricity came up with the idea, why not levy supermarket profits and then use the money so that each supermarket can become an autonomous correction unit in its own right - thus putting back goods and services into the community that it serves.

Hang on a sec I can hear you say? What villains in the checkouts, strange individuals cleaning the toilet and a dodgy lot on the loading bay – and you will all say, not in my backyard!

But wait a minute, how about all that wonderful black stuff that surrounds the supermarkets which are known as prairies in the summer as lack of rain causes a modern version of tumbleweed to be blown across the vast empty spaces when not in use, or at wind chill factor 15 icy winds turn it into a Siberian landscape.

What we need to do is to make it user friendly. Roof it over with high arches, art décor plaster and bright lights – and on top of the roof we put a modern version of the stocks which we will call correctional igloos.

Could one say one hundred modules per portion of the roofed over supermarket?

If this policy is adopted it will stop in its tracks society's vibrancy dying, "for every hardwood tree cut down, many more will die", is a very good metaphor for the current prognosis of not curing society's ills. Which daily fall into the category of apathy, as the spirit of individualism and imagination is sucked into a self destructive quagmire of petty legislation and (a metaphor for speed cameras set up in the workplace) to mask the demands of the voracious black hole of taxation. To feed the negative industry of crime generating, easily taxable jobs in parole monitoring, electronic tagging, probation and community service, and last but not least, the tried and tested blag of 'stay out of jail card – can't pay, won't pay, slap on the wrist, go away and pick up a caution some other day.

As our planet slowly dies, society has now the unfettered chainsaw armies or community delinquents who are let out of prison early to save money on one balance sheet only to see it flushed down the drain on another.

Hardworking individuals building a life around family and forming a strong community bond, for they strive against all

So You Want To Build the Naked UFO/Plasma Active Craft

odds to reach the bright sunlight, which is their canopy of spiritual and physical fulfilment. Loving, caring, tax generating economic units who would bring forth the next generation of like minded souls.

Cut down, mown down, murdered under the euphemism of manslaughter, lives selfishly squandered on the adrenaline of self-gratification of the ever growing minority as they trash the future of those left behind to mourn loved ones needlessly wasted on the roadside of the dark side of life. For without my correctional igloos, the punishment of mad, bad economics will never fit the crime.

Set Plasma Free (SPF) will give back to society its spirit of adventure for they once again can let down their drawbridges, open up their shutters, let in the sweet smell of reduced taxation, motivating them to save for the next great adventure, space – the last frontier – the ultimate dream, sponsored by grateful government that can take pride in the achievements of its citizens.

SPF will see the masses ascend into space and gaze in wonderment of the unique turquoise jewel, an oasis of mankind before the human race, as Earth's only parasite, will destroy it and leave it a salty red, no evidence left behind.

Let's Whip the Craft Into Shape (SPF)

Mankind has always looked for the easy solution to a problem, when in his early evolution the problem did not exist, until one day a small fire started by a lightning bolt gave him and some of his tribe some warmth.

Could his early brain be capable of computing the dry wood from the cave would help him move the fire from outside into the family domain? His problem now was how to keep the home fires burning, bring us right up to date.

Gas is a good friend, you can compress it and it will travel for miles across continents and even under seas.

The problem is all too easy. Now SPF will show how one man's evolved brain has done the impossible of turning one obnoxious, irritable liquid into a peaceful saviour of Mankind.

SPF will be the alchemist dream of converting base metals into gold as the unacceptable face of capitalism and wannabe industrial carpetbaggers, as asset strippers go forth with their battle cry "the sum of the pieces is greater than that of the whole"

Ladies and Gentlemen, please put your hands together for ammonia – nature's natural hydride. Hydride is a fancy word for sponge, but ammonia written H^3N tells those who have a desire and a need to understand the laws of the building blocks of the meaning of life, is a close cousin of water H^2O.

Water that makes the Earth the "turquoise jewel, an oasis of Mankind" is the life blood of civilisation. Wars have been fraught over it and one third of the Earth's population are deprived of it to a greater or lesser extent.

So You Want To Build the Naked UFO/Plasma Active Craft

Oh the water is there but the water runs deep, just dig a well and with the water irrigate your crops and let your goats drink. Would you take a pick and a shovel and dig a hole 160 feet deep and tap into this Third World liquid gold?

Ammonia is H^3N and water H^2O – you can see that H^2O has an O and not an N.

You could say that NO to H^3N and it is ON for H^2O.

SPF will deliver fresh water from seawater, which will end the poverty of the ironic arid coastal plains which are on or near the equator "where the sun keeps shining and the crops are failing as well". SPF takes us right back to our primitive roots with the first glimmer of an idea that fire can be kept alive by feeding it with dry wood, the oxygen carrier, the big O that breaths life into fire.

Today's problem – how much SPF must be used to separate the three atoms of hydrogen from the ammonia molecule and the two atoms of hydrogen from the water molecule and what to do with the "N" "O".

Back to the H^3N, which any mother will tell you is the hot rancid stench of urine – and if baby's bottom is not wiped and bathed to remove the residue, little baby will have a sore bum.

Let us stay in the nursery and move to a more underprivileged time where children were kids and could play a game up chimney stacks with brushes to see who could bring down the most soot without the brickwork of the stack cutting into their hands, or falling on their unprotected heads. Or even select the big lumps of coal from the smaller ones. This game was much more fun than today's finger painting – and you got paid as well.

Set Plasma Free

Those children who weren't game fully employed at the pithead or climbing chimneys or even crawling around under the weaving frames to pick up the cotton balls, could be seen to be playing with beautifully carved lumps of wood that looked like spinning tops. They say the children will learn from their elders, so it was not surprising that the top toy which was the hit of the day way back in Edwardian times was the whipping top. This insidious game of subjugation was to see who could use the whip to subjugate the top so that it would spin at high speed for a long time.

The inevitable happened, adults found a way to exploit this innocent play to put teams together and place bets on which team would win by making their spinning tops perform all sorts of gymnastic tricks.

It is said that behind closed doors the same whip was used for not such an innocent pastime but could be turned on the children who did not meet the trainer's expectations.

So You Want To Build the Naked UFO/Plasma Active Craft

What Would NASA Do With A Trashcan?

What can I do with strong standard traditional metal dustbins?

A characteristic of their design is to have been fluted all round. This is not only aesthetically pleasing but adds rigidity to the design so that a lighter and thinner metal can be used and for good measure the same concept is used on the lid. If I told you that a conjurer could pull out a reasonably sized kangaroo out of it then 610 cms in height and 460 cms in diameter would mean something.

As I am not a conjurer, nor do I have any aspiration to become one, all I shall pull out will be a shaft attached to the lid with a few magnetic bearings and gobbledegook thrown in for good measure. Attached to the lid, which has a series of thick copper discs. Interspaced between each disc is a pick up point, we will call it a brush as it does just that – it brushes the disc. "Each disc must cut a magnetic return circuit?" So there you see and now you don't – so now I have dropped it back in and pumped out all the air so that when it whirls round it will not get any indigestion as it is running in a vacuum.

Engineers and scientist who know about such things call this a homopolar gyroscopic generator (the gyroscopic indicates high inertia and each time current is withdrawn from the copper discs this will give you one pulse of angular momentum). "In simple terms it will give a kick like a 303 rifle that's not tucked tightly into your shoulder when you press the trigger".

All you and I need to know is that we have reproduced the storage ability in minutiae (this could well be the opposite of magnitude) of a thundercloud – then, as they say, a homopolar generator is the only game in town.

Set Plasma Free

I like the expression 'now you see it now you don't' for as the mist blows away we see our 30 foot diameter saucer, remember the one we made previously, supported on three equally spaced teacups.

A useful interpretation of a teacup is an empty drinking vessel, so we must fill each one with a high current (hence the need for thick copper discs), low voltage homopolar generator. Right in the vertical centre line we have a smaller teacup which because of its reduced size does not come in contact with the ground. In to this cup goes a fourth homopolar generator – this time the discs have increased in numbers but are far thinner in construction.

The thickness of the discs is proportional to the current it needs to convey. Our multitude of thin discs will therefore provide us with a very high voltage. This will be our equivalent to the thin blue lightning discharge which initiates the main current discharge back to its self. As this high current now has a high voltage ionised path to travel down and by clever use of discharge points, we can now whip round the high voltage in a manner similar to that of the Edwardian whipping tops!

As each electrical whip discharge produces an atmospheric plasma from cup A back to the high voltage homopolar generator D, therefore the next in line will 'B' and then 'C' and then back to 'A'.

In SPF my boast was that the plasma, once set free, would consume vast amounts of electrons. This will only happen if our SPF flying saucer is high enough off the ground so SPF will go on gorging itself without going to rest or asleep on the earth.

Now SPF although is only playing the ABC game, the revolving spark effect of the ABC game has bullied or

So You Want To Build the Naked UFO/Plasma Active Craft

whipped the ionised/plasmased air into a frenzy of a violent rotation, similar to that of the twister or tornado that lifted Dorothy's house to the Land of Oz.

As I talk of a tornado, let's go back to the previous picture of the Rolls Royce flying bedstead and you will immediately see that I have left out four military type axial jet engines, so in way of a good pun let's nick four from out of service redundant NATO Tornados.

Now they are all installed, once ignited they will blow vast amounts of very hot air at high pressure over our homopolar generators ABC and D with their dustbin lid like turbines so that most of the jet thrust will power the SPF craft to a height when the whole of SPF physics can come into play – and as it does so, we can truly say, that we have whipped the plasma to a state of subjugation so that it performs for us making SPF our personal slave to take us to Mars if not the stars!

A Wing and a Prayer

This book was never going to be easy to write, for no matter how I laced it with ironic humour, at the back of my mind were those horrifying visions captured, still to this day, as instant playback of the night of the 17th August 1943.

Was it all worth it? Only the publication of this book will put the knowledge that I paid such a high price for, when I accepted the responsibility of the papers which I took into my possession. Over the years I have cried many tears as I fought for inspiration to interpret the line drawings and the sketches that my father passed into my hands. Could it be that destiny has given me the responsibility of saving the human species from complete annihilation by events which are euphemistically called "Acts of God".

Before I move on, let me linger for a while on a wing and a prayer – as I confront the ghosts from the past and see it from the other side – for they were only doing their jobs.

With burning wings and fluttering tail planes, they fell on proud cities without remorse. Hamburg, Berlin and Schweinfurt and back again.

The shrunken corpses were just a token of B29s' incinerating wave that lit up the sky with the rising sun.

An atomic explosion, not a hydrogen bomb, a chain reaction of the neutron song, ended the evil that Man had begun.

Remember that Man, not gifted with politicians double speak; never, never and never again. Peace is with us, but the price was high – tens of thousands of men but boys shot out of the sky.

So You Want To Build the Naked UFO/Plasma Active Craft

Sixty three years of peace in Europe, with only a minor Balkan war which we brush aside as a bush fire - my apologies for the inappropriate pun.

Fifty million died for a quantum leap in consumerism. What a glib statement that is! Is it the aim of entrepreneurs to put space travel within reach of the average consumer?

Yes and I must play my part with Set Plasma Free available to all, there will be no more burning wings nor fluttering tail planes, for as I have emphasised over and over again, Set Plasma Free is one large circular wing and has no tail plane.

For those not familiar with how an aeroplane gets off the ground and flies and manoeuvres in the air, I will just enlighten them with a few words. As the plane races down the runway the air must travel much faster over the top of the wing, thus making the air at a lower pressure, to the bottom of the wing to the underside of the wing so the plane rises into the sky. Once in the air simply by disturbing this airflow by mechanical means the plane can be manoeuvred.

Set Plasma Free is no different to control than any of the planes that destroyed one man's nightmare and so gave birth to a dream.

Set Plasma Free will never fly by wire or be controlled by computers it must be flown by the seat of the pants, by direct mechanical linkage through a control column and even rudder pedals.

With such linkage we can open and close mechanical slats or slots which will run parallel to the mythical sides of our equilateral triangle.

What conclusions can one draw from this design feature? Would the electrons have a similar flow characteristic to water molecules and follow the most direct path to find their own level?

The above is purely a rhetorical question as the electrons will indeed cascade like a waterfall to reinforce the plasma nearest the slots. Any aeronautical engineer will tell you this is equivalent to blowing air over the wing or over one or other of the relevant control surfaces.

Why no computers? Why no wires? How do we light the inside of our Set Plasma Free spacecraft? The third question must answer the first two. With such a humongous flow of electrons an electro-magnetic field will be generated that would saturate and render useless any of the kit which we might have built in for the 20th Century, never mind the 21st Century. That's the bad news. The good news is that same electro- magnetic force (EMF) would be responsible for exciting any gas that sandwiched between two translucent sheets.

Will I be around to see fleets of Set Plasma Free craft, as numerous as wide bodied jets are today, embark on a trip in search of the sun taking the masses to the freshly terra formed Mars and beyond? Will Mankind build a new Jerusalem in such a green and pleasant land? I fear not – for Mankind must always be an interstellar parasite destroying everything, everywhere he goes.

Just a salty red, no evidence left behind.

www.ingramcontent.com/pod-product-compliance
Ingram Content Group UK Ltd.
Pitfield, Milton Keynes, MK11 3LW, UK
UKHW041413180426
11947UKWH00007B/106